TOUGH TRUCKS

TOYOTA™ TRUCKS

BY SETH LYNCH

Gareth Stevens PUBLISHING

HOT TOPICS

Please visit our website, www.garethstevens.com. For a free color catalog of all our high-quality books, call toll free 1-800-542-2595 or fax 1-877-542-2596.

Cataloging-in-Publication Data
Names: Lynch, Seth.
Title: Toyota trucks / Seth Lynch.
Description: New York : Gareth Stevens Publishing, 2019. | Series: Tough trucks | Includes glossary and index.
Identifiers: LCCN ISBN 9781538231562 (pbk.) | ISBN 9781538230329 (library bound) | ISBN 9781538231579 (6 pack)
Subjects: LCSH: Trucks--Juvenile literature. | Automobiles--Juvenile literature. | Toyota Jidōsha Kōgyō Kabushiki Kaisha--History.
Classification: LCC TL230.15 L96 2019 | DDC 629.224--dc23

Published in 2019 by
Gareth Stevens Publishing
111 East 14th Street, Suite 349
New York, NY 10003

Designer: Katelyn E. Reynolds
Editor: Kristen Nelson

Photo credits: Cover, p. 1 Yury Stroykin/Shutterstock.com; cover, pp. 1–32 (background) phiseksit/Shutterstock.com; p. 5 Darren Brode/Shutterstock.com; p. 7 Morio/Wikipedia.org; pp. 9, 30 (Model G1) Mytho88/Wikipedia.org; pp. 11 (inset), 30 (Hilux) Tennen-Gas/Wikipedia.org; p. 11 (main) Ovu0ng/Shutterstock.com; p. 13 JOHN LLOYD/Aabdullayev851/Wikipedia.org; p. 15 ShadowSlayer18/Wikipedia.org; pp. 17 (inset), 30 (2006 Tundra and 1998 Tacoma) IFCAR/Wikipedia.org; p. 17 (main) Philip Lange/Shutterstock.com; pp. 19, 30 (2018 Tacoma) Bilgin S. Sasmaz/Anadolu Agency/Getty Images; pp. 21, 30 (2018 Tundra) Darren Brode/Shutterstock.com; p. 22 marekuliasz/Shutterstock.com; p. 23 jrok (https://www.flickr.com/photos/jrok/)/Liftarn/Wikipedia.org; p. 24 Gavin Lawrence/Getty Images for NASCAR; p. 25 Jonathan Ferrey/Getty Images; p. 27 Nathan Pang/Shutterstock.com; p. 29 Bhakpong/Shutterstock.com; p. 30 (Stout) TTTNIS/Wikipedia.org.

Printed in the United States of America

CPSIA compliance information: Batch #CW19GS: For further information contact Gareth Stevens, New York, New York at 1-800-542-2595.

CONTENTS

SO MANY SOLD

Toyota Motor Corporation is one of the largest automobile **manufacturers** in the world. In 2016, Toyota sold more than 10 million **vehicles**! More than 190,000 of those were Toyota's mid-size truck, the Tacoma. About 115,000 were its full-size pickup, the Tundra!

FULL THROTTLE!

A pickup truck is a vehicle with a cab, or driver's area, that's connected in the back to a bed that's open and has low sides.

IT STARTED WITH LOOMS

Toyota has been making vehicles since the 1930s! The company began as part of Toyoda **Automatic** Loom Works, which manufactured tools called looms that made cloth. Toyota's founder, Sakichi Toyoda, wanted to challenge Ford and General Motors, companies that had begun selling cars in Japan.

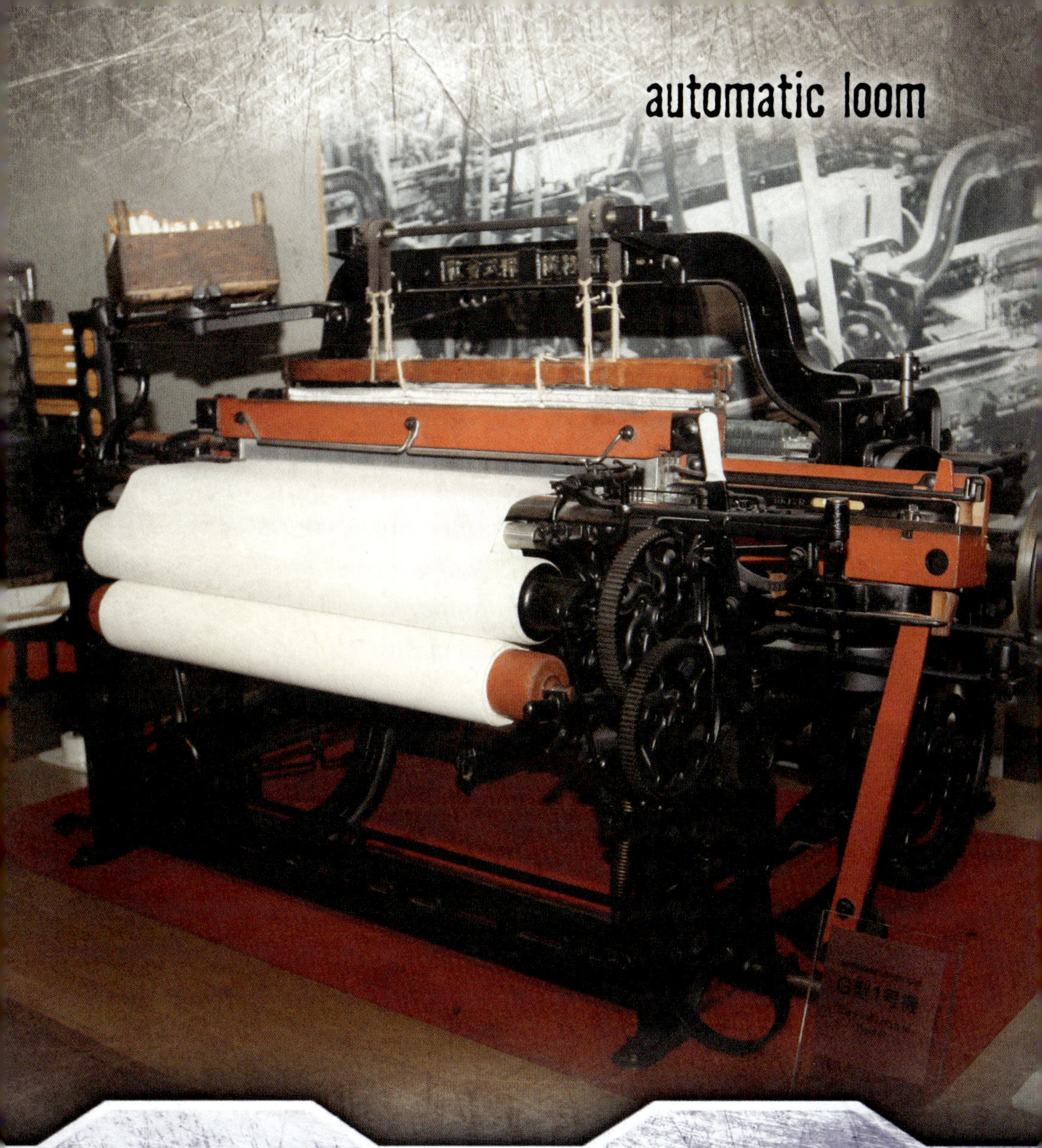

FULL THROTTLE!

Today, Toyota sells in about 190 countries and areas all over the world.

THE FIRST

The first vehicle the company put out was a truck! The Toyoda Model G1 began being produced in 1935. It looked a lot like the Ford trucks of that time. Only 379 of the Model G1 trucks were made.

replica of the Model G1

FULL THROTTLE!

The first Toyoda Model G1 truck was taken for a 783-mile (1,260 km) test drive through the Japanese cities of Tokyo, Gunma, Nagano, Yamanashi, and Hakone.

TRUCKS IN THE STATES

Toyota began selling vehicles in the United States in 1957. By the 1960s, it was selling trucks there, starting with the Land Cruiser and the Stout. In 1969, the Toyota Hilux was introduced in the United States.

first **generation** Hilux

2016 Hilux
(not sold in the US)

FULL THROTTLE!

Toyota made four-wheel drive vehicles for the Japanese Army before the start of World War II (1939–1945). A truck with four-wheel drive has an **engine** that supplies power to all four wheels.

The first Hiluxes had only 84 horsepower. The second-**generation** Hilux, which came out in 1973, increased to 109 horsepower. Toyota started calling these vehicles simply "Truck" instead of Hilux so buyers would have no question as to what the vehicle was.

FULL THROTTLE!

Horsepower is a unit used to measure the power of an engine.

The Toyota Truck produced in 1975 was bigger and had a new look. In 1979, the company added four-wheel drive, too. By the 1980s, the choices available for the Toyota Truck grew. Larger cabs, better engines, and automatic **transmissions** became possible options.

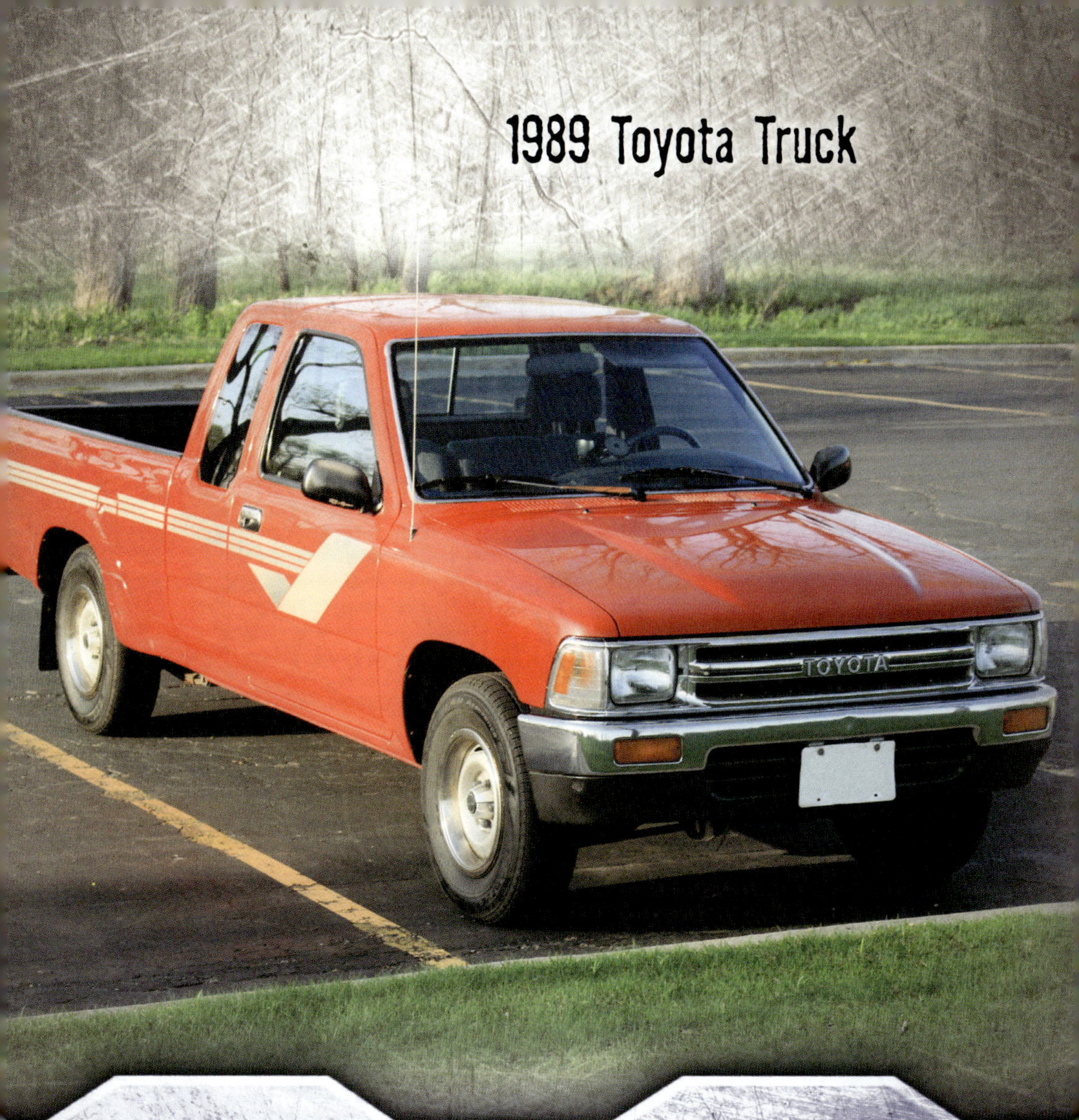

1989 Toyota Truck

FULL THROTTLE!

In 1989, the sixth generation of the Toyota Truck won *Motor Trend*'s 1989 Truck of the Year award.

NAMING THE TRUCK

New lines of the Toyota Truck continued coming out into the 1990s. In 1995, Toyota finally gave its mid-size pickup truck a name: the Toyota Tacoma. A full-size pickup, the Toyota Tundra, followed it on the road for the 2000 model year.

Tundra

Tacoma

FULL THROTTLE!

Toyota produced a mid-size pickup beginning in 1993 called the T100. Its small engine and size made it unpopular among most truck buyers, though!

THE TACOMA

Today, the Tacoma is one of the best-selling pickup trucks. The 2018 Tacoma can carry over 1,100 pounds (490 kg) and pull up to 6,400 pounds (2,902 kg)! It has a V6 engine and handles well even off-road.

2018 Tacoma TRD Pro

FULL THROTTLE!

A V6 is an engine with two banks of three **cylinders** in a V shape. V8 engines have two banks of four cylinders.

THE TUNDRA

The Tundra has a powerful V8 engine and it can tow a lot of weight. The 2018 Tundra Limited can pull more than 10,000 pounds (4,536 kg)! The Tundra comes in a number of trim levels that can include larger tires, heated seats, and special wheels!

2018 Tundra

FULL THROTTLE!

Trim levels are groupings of different features buyers can pay to add on when buying a vehicle.

BODY-ON-FRAME

Toyota makes several sport utility vehicles (SUVs), but the 4Runner is a bit different: It's a body-on-frame SUV. It looks like an SUV but is said to drive like a pickup! People who like to drive off-road often like the 4Runner.

2016 4Runner

2007 Tundra frame

FULL THROTTLE!

A body-on-frame vehicle is one that has a body separate from the frame. All cars used to be made this way, but today it's mostly trucks.

NASCAR TRUCK SERIES

Toyota has been racing its cars since 1957, but it wasn't until 2004 that it took its trucks to the track. In 2006, Toyota won the manufacturer's and driver's titles at the NASCAR Craftsman Truck Series!

FULL THROTTLE!

Driver Christopher Bell won the NASCAR Camping World Truck Series Championship driving a Tundra in 2017!

OFF-ROAD

In 1993 and 1998, a Toyota driver won the Baja 1000, a famously long off-road race. The company entered the 2015 TRD Pro Series Tundra in the race in 2014. It finished the 1,275-mile (2,052 km) race first in its class!

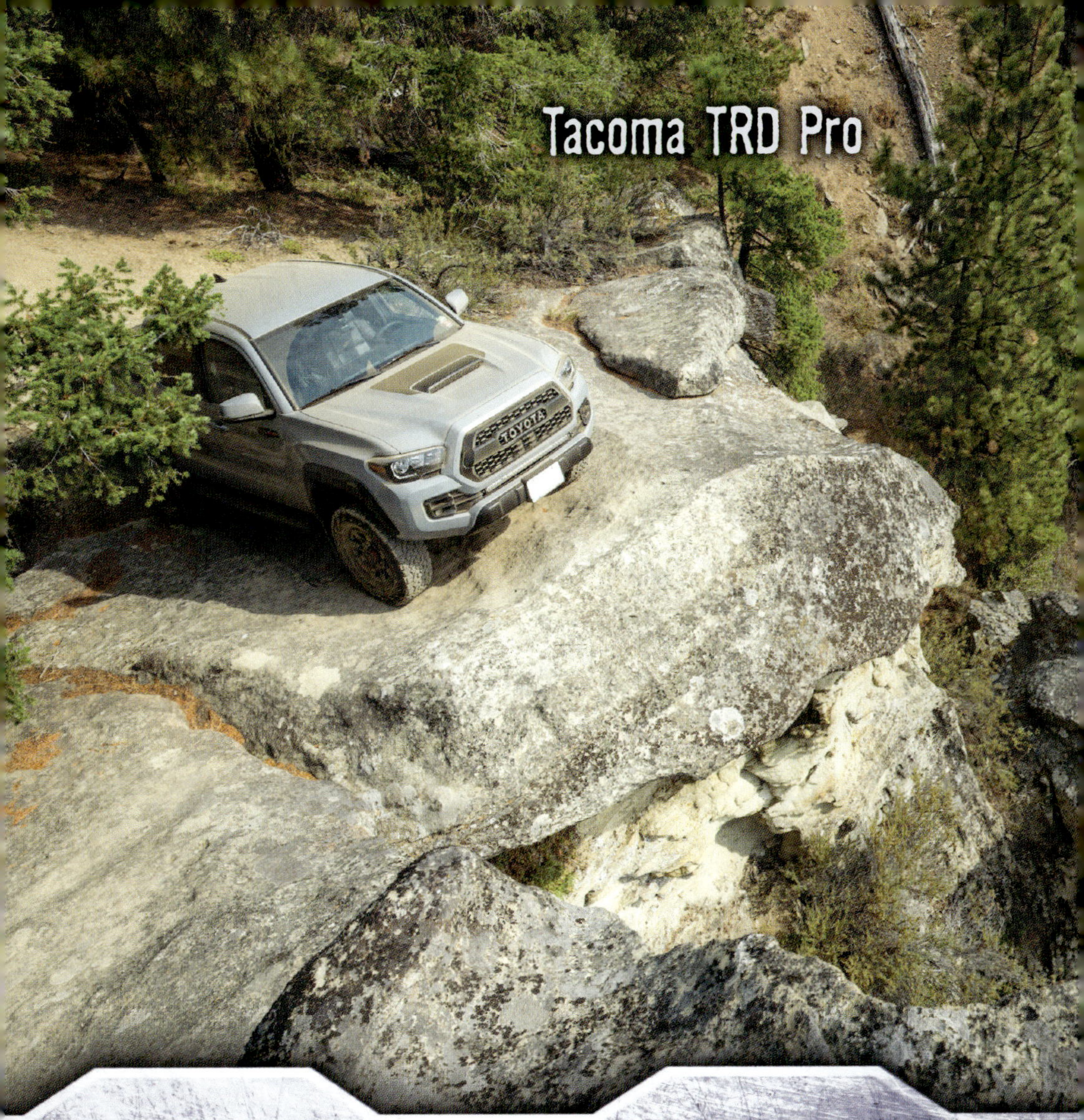

FULL THROTTLE!

Anyone can buy a TRD Pro Series Tundra, Tacoma, or 4Runner, not just racers!

WHAT'S NEXT?

Trucks continue to grow in popularity, and Toyota wants to continue working to make theirs stand out. From better gas mileage to ways to connect using a smartphone, Toyota is looking to the future with new looks and features for its trucks!

FULL THROTTLE!

Toyota makes some of the best-selling **hybrid** cars on the road. Will a hybrid truck be next?

TOYOTA TRUCKS THROUGH HISTORY

1935

Model G1

1960s

Stout

1970s

Hilux

1990s

Tacoma

2000s

Tundra

2018

Tacoma

Tundra

FOR MORE INFORMATION

BOOKS

Burgan, Michael. *Who Was Henry Ford?* New York, NY: Grosset & Dunlap, 2014.

Weber, M. *Wild Moments of Truck Racing*. North Mankato, MN: Edge Books, Capstone Press, 2018.

WEBSITES

Motor Trend
www.motortrend.com
Do you love cars and trucks? Read about them on *Motor Trend*'s website!

Toyota Official Website
www.toyota.com
Keep up to date with everything Toyota is doing on their official site!

Publisher's note to educators and parents: Our editors have carefully reviewed these websites to ensure that they are suitable for students. Many websites change frequently, however, and we cannot guarantee that a site's future contents will continue to meet our high standards of quality and educational value. Be advised that students should be closely supervised whenever they access the internet.

GLOSSARY

automatic: able to do work without a person at the controls. Also, a vehicle that changes gears by itself at different speeds.

cylinder: a tube-shaped space in an engine in which a piston moves up and down to create power

engine: a machine that makes power

generation: a group of things improved or developed from an earlier type

hybrid: able to run on electricity and gasoline but using one or the other at a time

manufacturer: a company that makes goods

transmission: the gear that sends power from a car or truck's engine to the wheels

vehicle: an object used for carrying or transporting people or goods, such as a car, truck, or airplane

INDEX